Chemistry
Required Practicals
Exam Practice Workbook

Primrose Kitten

GCSE

Contents

Introduction **3**

1	Making salts	**4**
2	Neutralisation	**8**
3	Electrolysis	**16**
4	Temperature changes	**22**
5	Rates of reaction	**28**
6	Chromatography	**36**
7	Identifying ions	**42**
8	Water purification	**48**

Periodic table **54**

Introduction

As part of your AQA GCSE Chemistry course, you will carry out eight Required Practicals. You can be asked about any aspect of any of these during the exams; this can include planning an investigation, making predictions, taking readings from equipment, analysing results, identifying patterns, drawing graphs, or suggesting improvements to the method. You can also be asked about practicals that are similar but that you may not have done before. You need to be able to recognise and apply the key practical skills that you have learnt to different experiments.

These practical questions account for at least 15% of the total marks. This Exam Practice Workbook allows you to practise answering questions on the eight Required Practicals and become familiar with the types of questions you may find in the exams. There are lots of hints and tips about what to look out for when answering practical questions.

Practical method – Full details of all eight Required Practicals, including equipment, method, and safety information, will remind you of the practical work you have carried out and the important skills you have gained during the course

Exam tips – Hints on how you can approach the practical exam questions, improve your answers, and secure marks

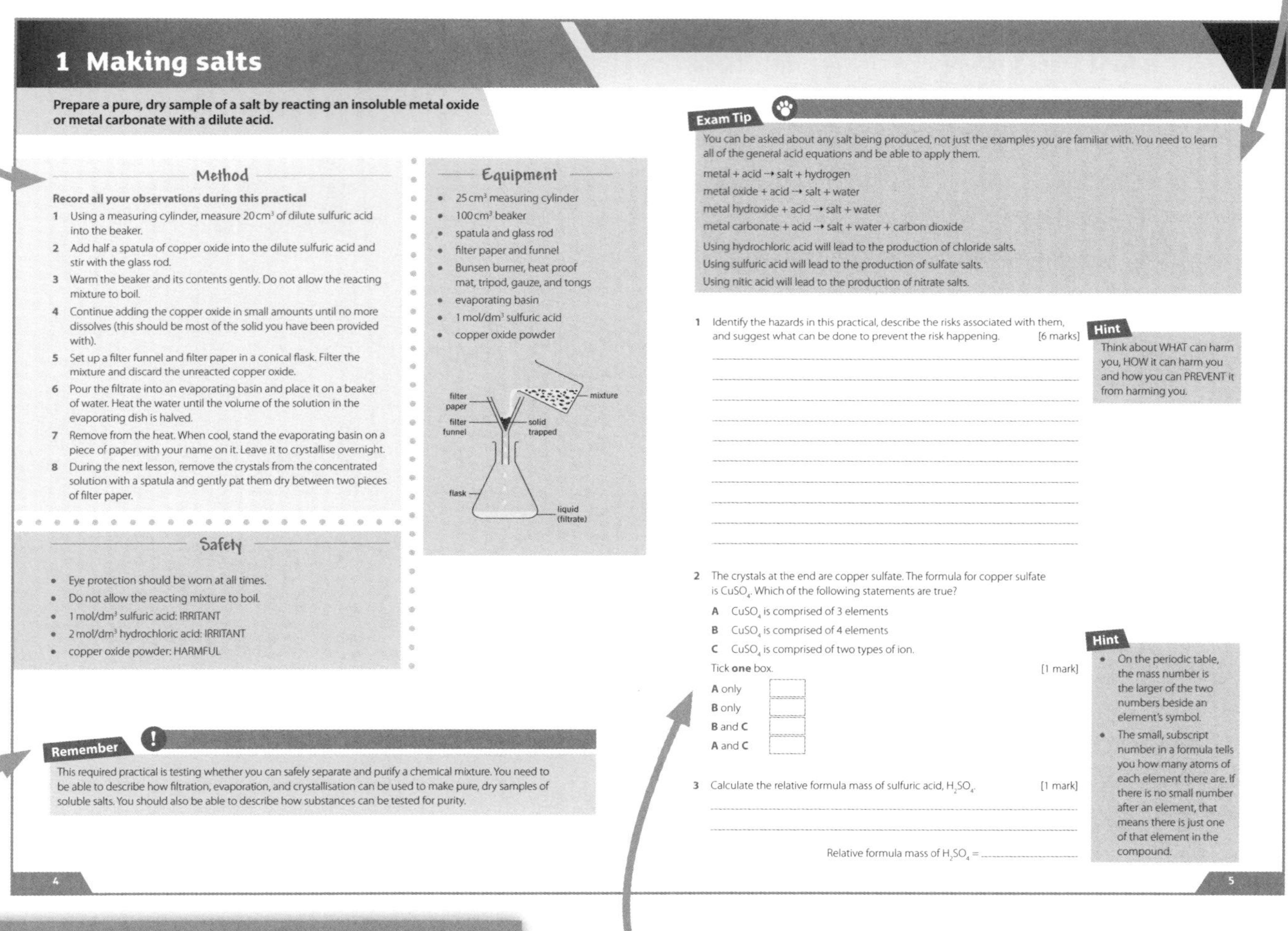

1 Making salts

Prepare a pure, dry sample of a salt by reacting an insoluble metal oxide or metal carbonate with a dilute acid.

Method

Record all your observations during this practical

1. Using a measuring cylinder, measure 20 cm^3 of dilute sulfuric acid into the beaker.
2. Add half a spatula of copper oxide into the dilute sulfuric acid and stir with the glass rod.
3. Warm the beaker and its contents gently. Do not allow the reacting mixture to boil.
4. Continue adding the copper oxide in small amounts until no more dissolves (this should be most of the solid you have been provided with).
5. Set up a filter funnel and filter paper in a conical flask. Filter the mixture and discard the unreacted copper oxide.
6. Pour the filtrate into an evaporating basin and place it on a beaker of water. Heat the water until the volume of the solution in the evaporating dish is halved.
7. Remove from the heat. When cool, stand the evaporating basin on a piece of paper with your name on it. Leave it to crystallise overnight.
8. During the next lesson, remove the crystals from the concentrated solution with a spatula and gently pat them dry between two pieces of filter paper.

Equipment

- 25 cm^3 measuring cylinder
- 100 cm^3 beaker
- spatula and glass rod
- filter paper and funnel
- Bunsen burner, heat proof mat, tripod, gauze, and tongs
- evaporating basin
- 1 mol/dm^3 sulfuric acid
- copper oxide powder

Safety

- Eye protection should be worn at all times.
- Do not allow the reacting mixture to boil.
- 1 mol/dm^3 sulfuric acid: IRRITANT
- 2 mol/dm^3 hydrochloric acid: IRRITANT
- copper oxide powder: HARMFUL

Remember

This required practical is testing whether you can safely separate and purify a chemical mixture. You need to be able to describe how filtration, evaporation, and crystallisation can be used to make pure, dry samples of soluble salts. You should also be able to describe how substances can be tested for purity.

Exam Tip

You can be asked about any salt being produced, not just the examples you are familiar with. You need to learn all of the general acid equations and be able to apply them.

metal + acid → salt + hydrogen
metal oxide + acid → salt + water
metal hydroxide + acid → salt + water
metal carbonate + acid → salt + water + carbon dioxide

Using hydrochloric acid will lead to the production of chloride salts.
Using sulfuric acid will lead to the production of sulfate salts.
Using nitic acid will lead to the production of nitrate salts.

1 Identify the hazards in this practical, describe the risks associated with them, and suggest what can be done to prevent the risk happening. [6 marks]

Hint
Think about WHAT can harm you, HOW it can harm you and how you can PREVENT it from harming you.

2 The crystals at the end are copper sulfate. The formula for copper sulfate is $CuSO_4$. Which of the following statements are true?

A $CuSO_4$ is comprised of 3 elements
B $CuSO_4$ is comprised of 4 elements
C $CuSO_4$ is comprised of two types of ion.

Tick **one** box. [1 mark]

A only ☐
B only ☐
B and **C** ☐
A and **C** ☐

Hint
- On the periodic table, the mass number is the larger of the two numbers beside an element's symbol.
- The small, subscript number in a formula tells you how many atoms of each element there are. If there is no small number after an element, that means there is just one of that element in the compound.

3 Calculate the relative formula mass of sulfuric acid, H_2SO_4. [1 mark]

Relative formula mass of H_2SO_4 = ____________

4 5

Remember – Each practical has a reminder of the key skills being tested in the practical, whatever the context

Exam-style questions – Lots of practical exam-style questions about each Required Practical help you to become confident in answering practical questions

1 Making salts

Prepare a pure, dry sample of a salt by reacting an insoluble metal oxide or metal carbonate with a dilute acid.

Method

Record all your observations during this practical

1. Using a measuring cylinder, measure 20 cm^3 of dilute sulfuric acid into the beaker.
2. Add half a spatula of copper oxide into the dilute sulfuric acid and stir with the glass rod.
3. Warm the beaker and its contents gently. Do not allow the reacting mixture to boil.
4. Continue adding the copper oxide in small amounts until no more dissolves (this should be most of the solid you have been provided with).
5. Set up a filter funnel and filter paper in a conical flask. Filter the mixture and discard the unreacted copper oxide.
6. Pour the filtrate into an evaporating basin and place it on a beaker of water. Heat the water until the volume of the solution in the evaporating dish is halved.
7. Remove from the heat. When cool, stand the evaporating basin on a piece of paper with your name on it. Leave it to crystallise overnight.
8. During the next lesson, remove the crystals from the concentrated solution with a spatula and gently pat them dry between two pieces of filter paper.

Equipment

- 25 cm^3 measuring cylinder
- 100 cm^3 beaker
- spatula and glass rod
- filter paper and funnel
- Bunsen burner, heat proof mat, tripod, gauze, and tongs
- evaporating basin
- 1 mol/dm^3 sulfuric acid
- copper oxide powder

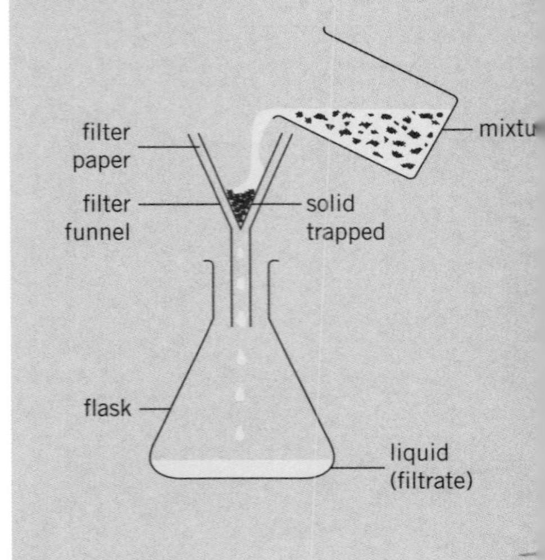

Safety

- Eye protection should be worn at all times.
- Do not allow the reacting mixture to boil.
- 1 mol/dm^3 sulfuric acid: IRRITANT
- 2 mol/dm^3 hydrochloric acid: IRRITANT
- copper oxide powder: HARMFUL

Remember !

This required practical is testing whether you can safely separate and purify a chemical mixture. You need to be able to describe how filtration, evaporation, and crystallisation can be used to make pure, dry samples of soluble salts. You should also be able to describe how substances can be tested for purity.

Exam Tip

You can be asked about any salt being produced, not just the examples you are familiar with. You need to learn all of the general acid equations and be able to apply them.

metal + acid → salt + hydrogen

metal oxide + acid → salt + water

metal hydroxide + acid → salt + water

metal carbonate + acid → salt + water + carbon dioxide

Using hydrochloric acid will lead to the production of chloride salts.

Using sulfuric acid will lead to the production of sulfate salts.

Using nitic acid will lead to the production of nitrate salts.

1 Identify the hazards in this practical, describe the risks associated with them, and suggest what can be done to prevent the risk happening. [6 marks]

The acids used are irritants. They may accidentally splash during pouring and go into eyes. To prevent this goggles should be worn. The flame to heat up the filtrate will be hot. The bunsen burner should be kept out of easy reach and on safety flame when not heating the filtrate. Copper oxide powder can be harmful so shouldn't be touched - wear gloves or use spatula

glass - break - cuts.

Hint

Think about WHAT can harm you, HOW it can harm you and how you can PREVENT it from harming you.

2 The crystals at the end are copper sulfate. The formula for copper sulfate is $CuSO_4$. Which of the following statements are true?

A $CuSO_4$ is comprised of 3 elements

B $CuSO_4$ is comprised of 4 elements

C $CuSO_4$ is comprised of two types of ion.

Tick **one** box. [1 mark]

A only ☐

B only ☐

B and **C** ☐

A and **C** ☑

Hint

- On the periodic table, the mass number is the larger of the two numbers beside an element's symbol.
- The small, subscript number in a formula tells you how many atoms of each element there are. If there is no small number after an element, that means there is just one of that element in the compound.

3 Calculate the relative formula mass of sulfuric acid, H_2SO_4. [1 mark]

Relative formula mass of H_2SO_4 = ____________

4 Suggest the function of the filter paper. [1 mark]

to remove the insoluable residue before evaporating.

Hint
What was left in the filter paper once you were finished with it?

5 Draw one line from each key term to its definition. [2 marks]

Key term	Definition
Filtrate	An insoluble solid formed by a reaction taking place in a solution.
Precipitate	A solution that has passed through a filter.
Salt	A compound formed when the hydrogen in an acid is replaced by a metal.

6 The method described for this practical evaporates the water from the solution by putting the evaporating basin on a water bath.

Suggest **one** other method of evaporating the liquid. [1 mark]

____

7 Describe the purpose of evaporating the water. [1 mark]

____

8 A student carries out a similar experiment with copper oxide and sulfuric acid and makes two observations.

- The black powered appears to 'disappear'.
- The solution turns from colourless to blue.

Explain why these changes happened. [4 marks]

____

9 Complete and balance the symbol equation for the reaction of copper(II) oxide with sulfuric acid. [2 marks]

CuO + H_2SO_4 → $CuSO_4$ + H_2O

10 The formula of copper(II) oxide is CuO.

Write the formula of copper(I) oxide. [1 mark]

____

11 Hydrochloric acid can be reacted with magnesium or magnesium carbonate. Both give magnesium chloride as one product and a gas as another product.

magnesium + hydrochloric acid → magnesium chloride + gas A

magnesium carbonate + hydrochloric acid → magnesium chloride + water + gas B

Identify gas A and gas B and give the test to confirm the identity of each gas. [4 marks]

Gas A is ______

Test for gas A:

______

Gas B is ______

Test for gas B:

______

12 A salt is a compound formed when an acid reacts with a base. Write word equations for the production of the following salts.

a The production of potassium chloride [2 marks]

______

b The production of iron sulfate [2 marks]

______

c The production of lead nitrate [2 marks]

______

Exam Tip

It is important that you can apply any of the general acid equations.

13 Insoluble calcium sulfate is produced in the reaction below.

$H_2SO_4(aq) + Ca(OH)_2(aq) \rightarrow CaSO_4(s) + 2H_2O(l)$

Explain what you would observe if you carried out this reaction. [2 marks]

______

______

Hint

Look carefully at the state symbols. Copper sulfate is a soluble salt, shown by the state symbol (aq). However, calcium sulfate is an insoluble salt and in the equation, we show this by using the state symbol (s).

14 Suggest why it is important that the copper oxide is in excess in this practical. [1 mark]

______

______

15 Ⓗ Calculate how much copper sulfate can be produced from 17.0 g of copper oxide.

Give your answer to three significant figures. [3 marks]

______

______

______

______

______

Mass of copper sulfate produced = ______ g

2 Neutralisation

Use titration to determine the volume of an acid needed to completely react with an alkali

Method

1. Collect some dilute sulfuric acid in a labelled beaker.
2. Fill a burette with the dilute sulfuric acid just beyond the zero mark, and then let the solution run out until the bottom of the meniscus is exactly on the zero mark.
3. Collect some sodium hydroxide in another labelled beaker.
4. Use the 25.0 cm^3 pipette and pipette filler to transfer 25.0 cm^3 of the sodium hydroxide into a clean, dry conical flask.
5. Add three to four drops of phenolphthalein indicator into the flask and swirl. Place the conical flask on the white tile directly below the burette.
6. Record the initial burette reading in a suitable results table.
7. Carry out a rough titration by adding the acid to the alkali in small amounts at a time. Swirl the flask after every addition and continue until the indicator changes from colourless to pink. Record the final burette reading in your results table.
8. Repeat the titration accurately by adding the acid drop-by-drop when you are near the end point.
9. Repeat the accurate titrations until you have two concordant results (within 0.1 cm^3 of each other).

Safety

- Eye protection must be worn
- 0.100 mol/dm^3 sodium hydroxide solution: IRRITANT
- dilute sulfuric acid: IRRITANT

Equipment

- 50 cm^3 burette, stand, and burette holder
- 2 × 250 cm^3 beakers
- 250 cm^3 conical flask
- 25.0 cm^3 bulb pipette and pipette filler
- white tile
- funnel
- phenolphthalein indicator
- 0.100 mol/dm^3 sodium hydroxide solution
- dilute sulfuric acid of unknown concentration
- wash bottle of distilled water

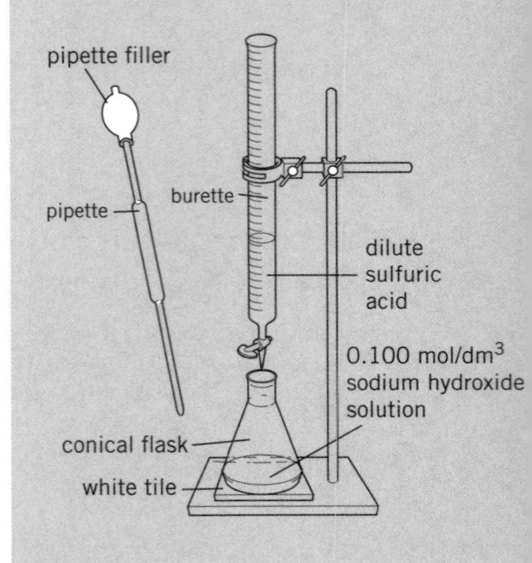

Remember !

This practical tests your ability to determine the concentration of a solution of acid or alkali. You should be able to describe how to use burettes and pipettes to accurately measure and transfer volumes of liquids.

Exam Tip

Foundation students need to know how to find the volume that will neutralise a solution. Higher tier students need to then use that volume to calculate the concentration of the solution.

A common question is to ask you to write a method for a titration. Make sure you spell the words burette and pipette correctly.

1 **a** The titration described in the method involves adding sulfuric acid to sodium hydroxide.

Name the type of reaction that occurs in the conical flask. [1 mark]

neutralisation

b Circle the ion that is produced by all acids in aqueous solutions. [1 mark]

H^+ Na^+ OH^- SO_4^{2-}

c Which of the following values is the pH you would expect a solution containing OH^- ions to be?

Tick **one** box. [1 mark]

- [ ] pH 3
- [ ] pH 5
- [ ] pH 7
- [x] pH 9

d Describe what concordant results are. [2 marks]

within 0.1 of each other

2 **a** Describe how you would safely fill a burette with sulfuric acid. [4 marks]

b Identify **two** possible sources of error in this experiment. [2 marks]

not equal volumes

parallax / meniscus

---

3 The diagram below shows a burette before a student started their titration.

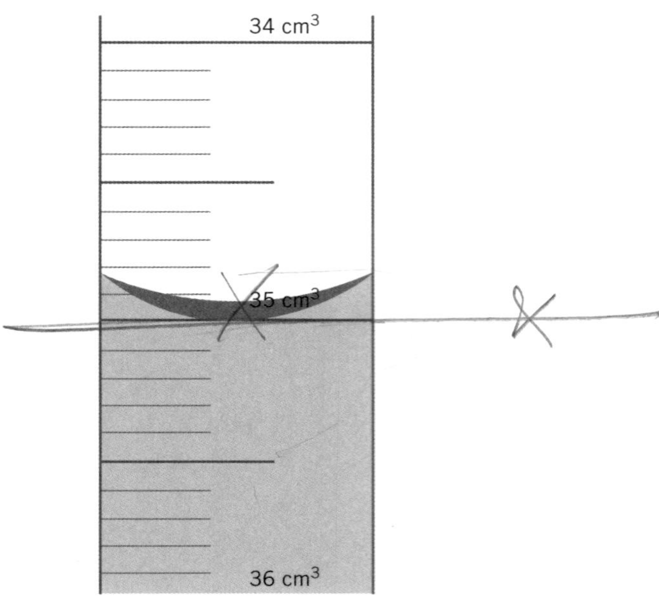

a Draw an 'X' on the diagram to show where the student should position their eyes when measuring the initial burette reading. [1 mark]

The diagram below shows the burette at the point when the solution in the conical flask turned pink.

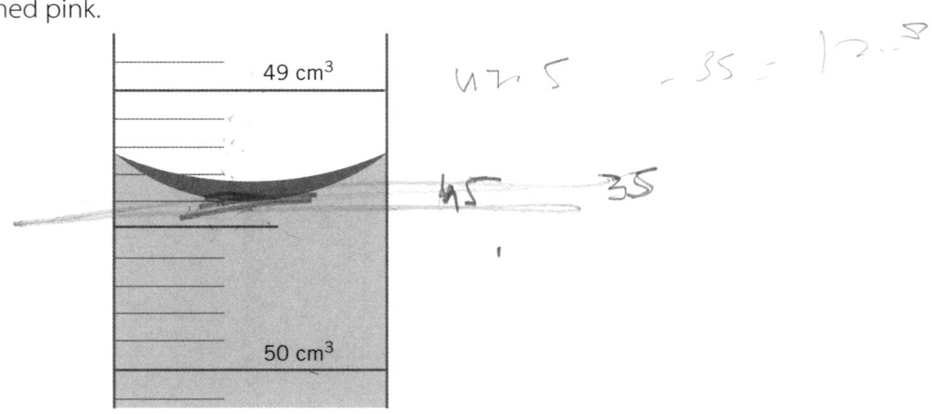

b Give the volume of dilute sulfuric acid that was required to neutralise the sodium hydroxide solution. [1 mark]

Answer = ______ cm^3

Hint

You need to subtract!

c Give your answer from **b** in dm^3. [1 mark]

Answer = ________ dm^3

4 In class a student carried out a titration between sodium hydroxide (NaOH) and sulfuric acid (H_2SO_4).

a Complete the word equation for this reaction. [2 marks]

sodium hydroxide + ________ → sodium sulfate + ________

b Give the balanced symbol equation for this reaction. [3 marks]

c **H** Give the number of moles of alkali that react with one mole of acid. [1 mark]

d **H** The student found that $13.6\,cm^3$ of $0.10\,mol/dm^3$ sulfuric acid was required to neutralise $25\,cm^3$ sodium hydroxide solution.

Calculate the concentration of the sodium hydroxide solution. [6 marks]

Concentration of NaOH solution = ________ mol/dm^3

Exam Tip

Make sure you know the general equation for the reaction between an acid and a base.

5 A student wanted to obtain sodium nitrate ($NaNO_3$) as a pure salt. They carried out a titration with nitric acid (HNO_3) and sodium hydroxide (NaOH).

a Calculate the atom economy of this reaction. [5 marks]

Atom economy = ______%

Exam Tip

This is a good example of how you could be asked to apply your maths skills to a practical context.

Don't be put off by having to draw on your knowledge of other parts of the Chemistry course.

The student's results are shown below.

Volume of dilute nitric acid added in cm^3			
Trial 1 (rough)	**Trial 2**	**Trial 3**	**Mean**
17.10	16.35	16.30	

b Explain the purpose of carrying out a 'rough' first titration. [1 mark]

c Calculate the mean volume of dilute nitric acid needed to neutralise the sodium hydroxide. [1 mark]

Mean titre volume = ______ cm^3

6 The 'end point' of a titration is when the acid or alkali in the conical flask has been neutralised.

a Describe how you can use an indicator to tell when the end point has been reached. [2 marks]

b There are two common indicators used in acid-base titrations:

- phenolphthalein indicator
- methyl orange

Complete the table below to show the colours each indicator produces in acidic and alkaline conditions. [2 marks]

	Colour in acid	**Colour in alkali**
phenolphthalein		
methyl-orange		

7 A student uses the equipment below to carry out a titration between sodium hydroxide and hydrochloric acid.

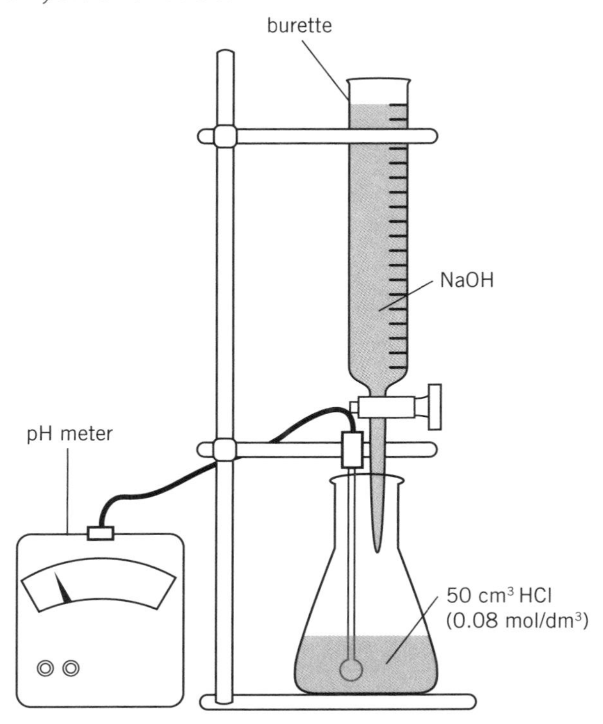

a Explain why a white tile is not needed in this version of the experiment. [2 marks]

b Describe how the student should use the equipment to carry out the titration. You can assume that:

- the burette is already filled with sodium hydroxide solution of unknown concentration
- the conical flask has already been filled with 50 cm^3 of 0.08 mol/dm^3 hydrochloric acid. [6 marks]

Hint

Look at the diagram carefully. In this case it is the alkali's concentration that the student is trying to find out. Be aware that it does not always have to be the acid in the burette. The theory behind the titration is exactly the same either way.

The student's results are shown in the graph below.

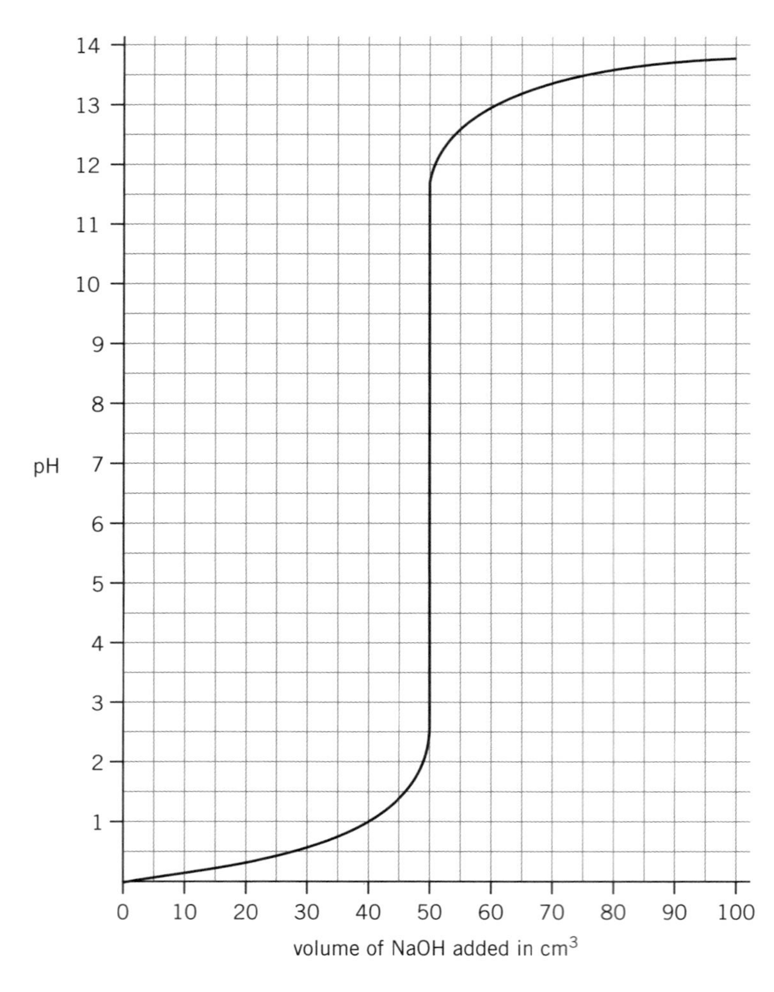

c The student concludes that $50\,cm^3$ sodium hydroxide was required to neutralise the hydrochloric acid.

Give evidence from the graph that supports the student's conclusion. [1 mark]

____

____

____

Hint

There is one mark available so you will only be expected to give one piece of information.

d (H) Calculate the concentration of the sodium hydroxide solution. [5 marks]

____

____

____

____

____

____

Concentration of NaOH solution = __________ mol/dm^3

e (H) Calculate the mass of hydrochloric acid that reacted with sodium hydroxide in this titration. [2 marks]

____

____

____

Answer = ____________ g

8 (H) Hydrochloric acid is a strong acid.

Explain why the pH of a strong acid is lower than the pH of a weak acid at the same concentration. [2 marks]

____

____

____

____

3 Electrolysis

Investigate the decomposition of two ionic solutions using electrolysis.

Method

1. Transfer 50 cm^3 copper(II) chloride solution to the 100 cm^3 beaker.
2. Insert the carbon electrodes into the solution, ensuring they do not touch.
3. Attach the electrodes to the DC power supply using the leads and crocodile clips.
4. Switch the power supply on at 4V and carefully observe what happens at the anode and the cathode. Record your observations.
5. Position a piece of damp litmus paper above the solution next to the anode. Record your observations.
6. Collect a new set of electrodes, wash the equipment, and repeat steps 1–5 with 50 cm^3 sodium chloride solution in the beaker.

Safety

- copper(II) chloride solution – IRRITANT
- oxygen gas – OXIDISING
- hydrogen gas – EXREMELY FLAMMABLE
- chlorine gas – TOXIC
- The electrolysis of brine produces a solution of sodium hydroxide, which is corrosive.
- Wear chemical splash proof eye protection.
- Wear nitrile gloves and only complete the practical in a well-ventilated room. Take extra care if you are asthmatic.
- Switch off the electric current as soon as you have made your observations.

Equipment

- 100 cm^3 beaker
- 2 × carbon electrodes
- 2 × crocodile clips and wires
- 1 × low voltage lab pack
- copper(II) chloride solution
- saturated sodium chloride solution
- litmus paper
- forceps
- eye protection and nitrile gloves

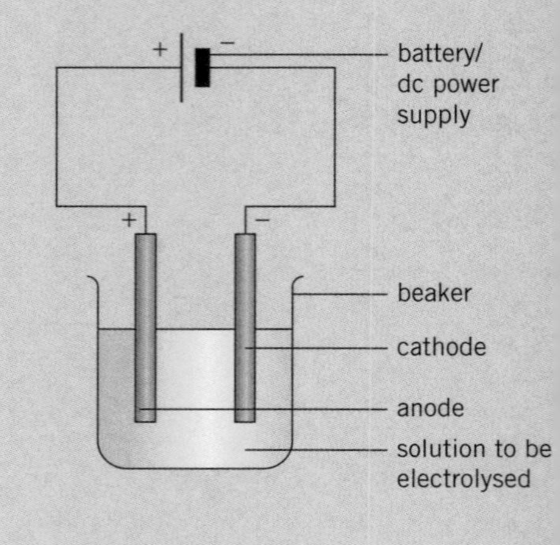

Remember !

This practical tests your ability to identify elements and compounds from observations. Remember that electrolysis uses electricity to break ionic compounds down into elements or simpler compounds. Metals or hydrogen are made at the negative electrode and non-metal molecules, including oxygen, are made at the positive electrode.

Exam Tip

There are many solutions that can undergo electrolysis. You need to be able to apply the principles of the practical you carried out to any example, including molten ionic compounds.

You should be able to predict the products of the electrolysis of different solutions and identify which ion goes to each electrode.

For higher tier, you should be able to write an equation for the reaction at each electrode.

1 Give the meaning of electrolysis. [2 marks]

____

____

2 Electrolysis involves the movement of ions towards electrodes.

Describe the difference between atoms and ions. [2 marks]

____

____

____

3 Draw one line from each statement to the electrode it describes. [2 marks]

Statement	Electrode
Positive electrode	
Negative electrode	Anode
Positively charged ions move towards it	Cathode
Negatively charged ions move towards it	

4 a Explain why electrolysis would not occur if the electrodes were touching each other. [2 marks]

____

____

____

b A student checks that the electrodes are not touching but still cannot see gas being produced at the electrodes.

Suggest an alteration to the experiment that would immediately show if the circuit was working properly. [1 mark]

____

____

5 Describe the hazards and/or risks associated with electrolysis of copper(II) chloride and the safety precautions that should be taken to reduce these risks. [6 marks]

6 Two students carried out electrolysis of copper(II) chloride.

- Student A set their power supply to 4 V.
- Student B used a power supply set at 1 V.

Their methods were the same in all other ways.

Compare the results you would expect from the two experiments. [4 marks]

Exam Tip

When 'compare' is the command word, you need to include things that are the same and things that are different.

7 In the two experiments described in the method, the copper ions and the sodium ions will both move to the same electrode.

Identify which electrode the copper ions and sodium ions will move towards and explain why. [2 marks]

Electrode = ______________________________

Explanation = ______________________________

8 The equipment in the diagram below is used to collect the gases produced during the electrolysis of calcium nitrate. [4 marks]

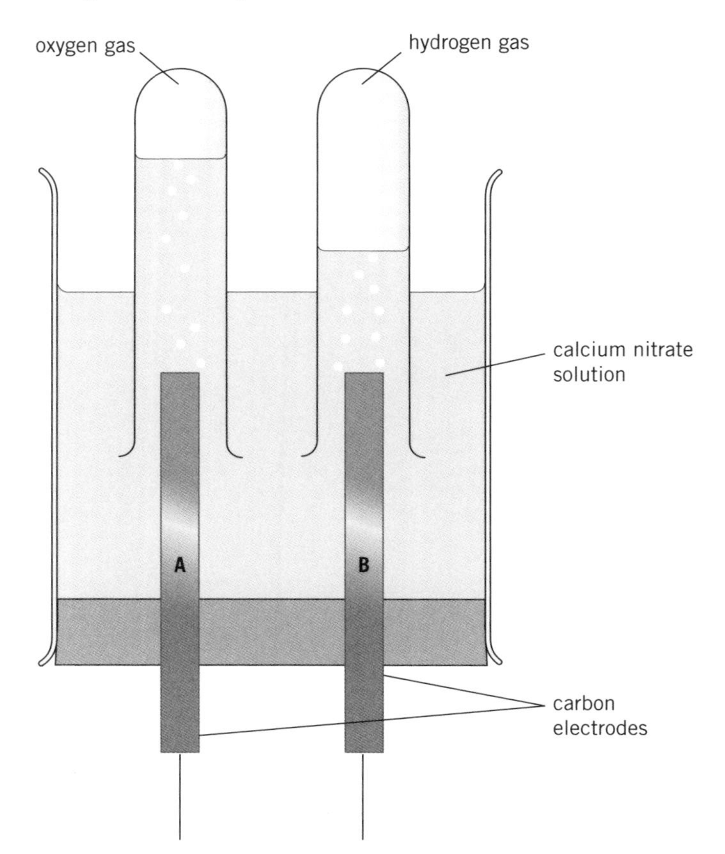

a Describe the tests you could carry out to confirm the identity of the two gases. [3 marks]

____

____

____

____

____

____

b Give the letter of the anode and the cathode. [2 marks]

Anode = __________

Cathode = __________

Hint

Think about the charges on each of the ions.

9 a In the electrolysis of sodium chloride solution, hydrogen gas is formed at one of the electrodes.

Give the reason why hydrogen gas is formed instead of sodium metal.

______

______

b There are three products from the electrolysis of sodium chloride; two gases and a compound in solution.

Give the formula of the compound in the solution. [1 mark]

Formula = ______

c Explain why solid sodium chloride cannot undergo electrolysis. [2 marks]

______

______

______

______

d Explain how and why you would expect the pH of the solution to change during sodium chloride electrolysis. [3 marks]

______

______

______

______

10 Ⓗ Use ionic half equations to describe what happens at each electrode in copper(II) chloride electrolysis. [6 marks]

______

______

______

______

______

______

______

______

______

______

Hint

Look at the number of marks available. Six marks doesn't mean you need to write an essay, but you do need to write three things about each electrode.

11 Ⓗ A teacher demonstrates the electrolysis of lead bromide. Their experiment is shown in the diagram below.

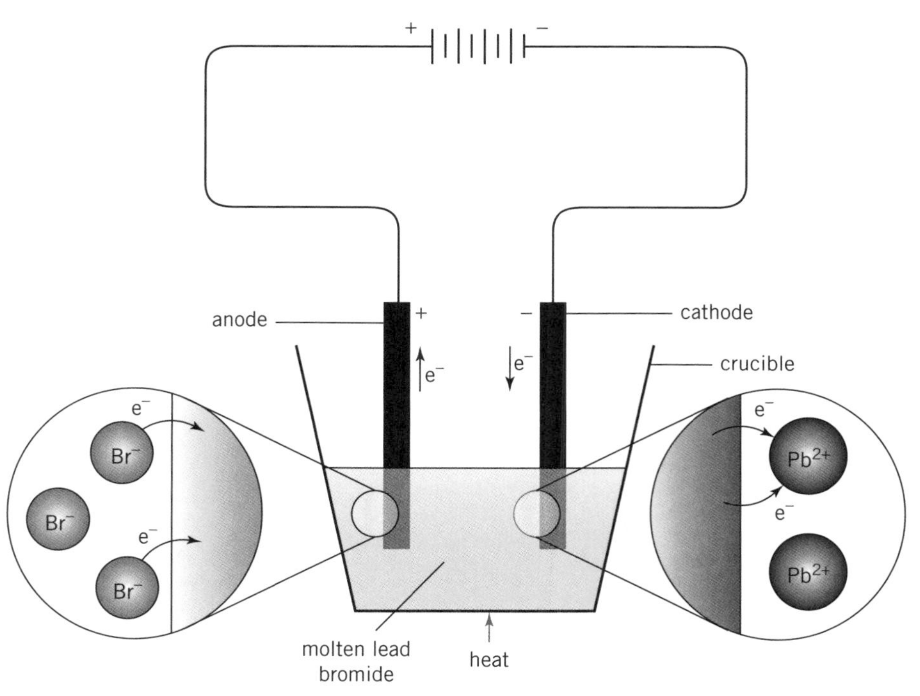

a Explain what will happen at each electrode in terms of oxidation and reduction. [4 marks]

b Suggest a safety precaution that should be taken while the teacher carries out the practical. [1 mark]

4 Temperature changes

Investigate variables that affect temperature change in reacting solutions

Method

1. Place the polystyrene cup inside a 250 cm^3 beaker.
2. Measure 30 cm^3 of dilute hydrochloric acid in the measuring cylinder and pour into the polystyrene cup.
3. Place the lid on the cup and insert the thermometer.
4. Record the temperature of the dilute hydrochloric acid.
5. Measure 5 cm^3 of sodium hydroxide solution, pour it into the cup, and stir to mix.
6. Record the highest temperature reached by the thermometer.
7. Repeat steps 5 and 6, adding 5 cm^3 sodium hydroxide solution up to a maximum of 40 cm^3.
8. Repeat the entire experiment another two times.

Safety

- Sodium hydroxide solution: CORROSIVE
- Hydrochloric acid: IRRITANT
- Wear chemical splash-proof eye protection and wash hands after the practical.

Equipment

- eye protection
- 50 cm^3 measuring cylinders and a 250 cm^3 beaker
- polystyrene cup and lid with a hole for a thermometer.
- weighing boat, spatula, and balance
- 0–110°C thermometer
- stopwatch
- dilute hydrochloric acid
- sodium hydroxide solution

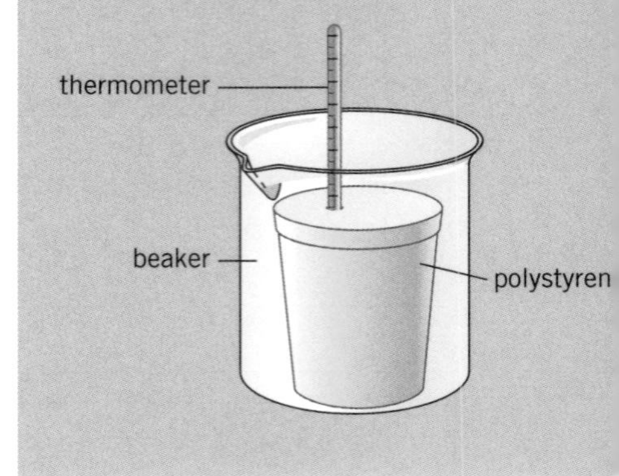

Remember

This practical tests your ability to safely and accurately measure mass, temperature, and volume in order to investigate chemical reactions. In this reaction you are mixing a strong acid with a strong alkali, as in Required Practical 8 Making salts. A key skill being tested is your ability to extract information from graphs.

Exam Tip 

You will need to know:

- the general equations for the reactions of acids and be able to apply them
- how to determine the formula of ionic compounds from the charges on their ions
- formulae of all the ions involved in the neutralisation reaction.

1 **a** Give a reason for using a cup made from polystyrene. [1 mark]

____

____

b Give the function of the beaker. [1 mark]

____

____

c Give the function of the lid. [1 mark]

____

____

2 Give two possible sources of error in this experiment. [2 marks]

____

____

3 Complete the balanced equation for the reaction of sodium hydroxide solution (NaOH) with dilute hydrochloric acid (HCl). [2 marks]

________ (aq) + HCl(aq) $\longrightarrow$ ________ (aq) + ________ (l)

Hint

Pay close attention to ALL the information being given to you in the equation.

4 Explain why it is important to wait until the reading on the thermometer stops changing before recording the temperature. [2 marks]

____

____

____

____

5 It is important not to leave long gaps between adding each 5 cm^3 sample of sodium hydroxide solution in this experiment.

Explain what effect adding the sodium hydroxide solution after a very long interval would have on the results. [2 marks]

____

____

____

6 The equipment must be washed thoroughly before repeating the whole experiment.

Why it is important to repeat the whole experiment?

Tick **two** boxes. [2 marks]

Calculate a mean	☐
Improve accuracy	☐
Reduce effect of random errors	☐
Spot anomalies	☐
Test reproducibility	☐

Exam Tip

Always double check how many boxes you are being asked to tick.

7 Explain briefly why it is important to wash out the equipment before you repeat the experiment. [2 marks]

____

____

8 a The reaction between sodium hydroxide and hydrochloric acid is exothermic.

Sketch a reaction profile for this reaction on the axes below. [2 marks]

energy

progress of reaction

Exam Tip

Remember that if you are asked to sketch a graph, you don't need to plot any points. Just show the correct overall shape of the graph.

b Describe what happens to the energy transferred in this reaction. [1 mark]

____

____

c Give another example of an exothermic reaction. [1 mark]

____

9 A student carried out an experiment to measure the energy change calculated from the results was +41.7 kJ/mol.

Describe what this tells us about the energy change in the reaction. [1 mark]

____

____

10 a Plot the given data below on the axes provided. [3 marks]

Volume of NaOH added in cm^3	Mean highest temperature reached in °C
0	19
5	22
10	25
15	27
20	29
25	31
30	34
35	33
40	30

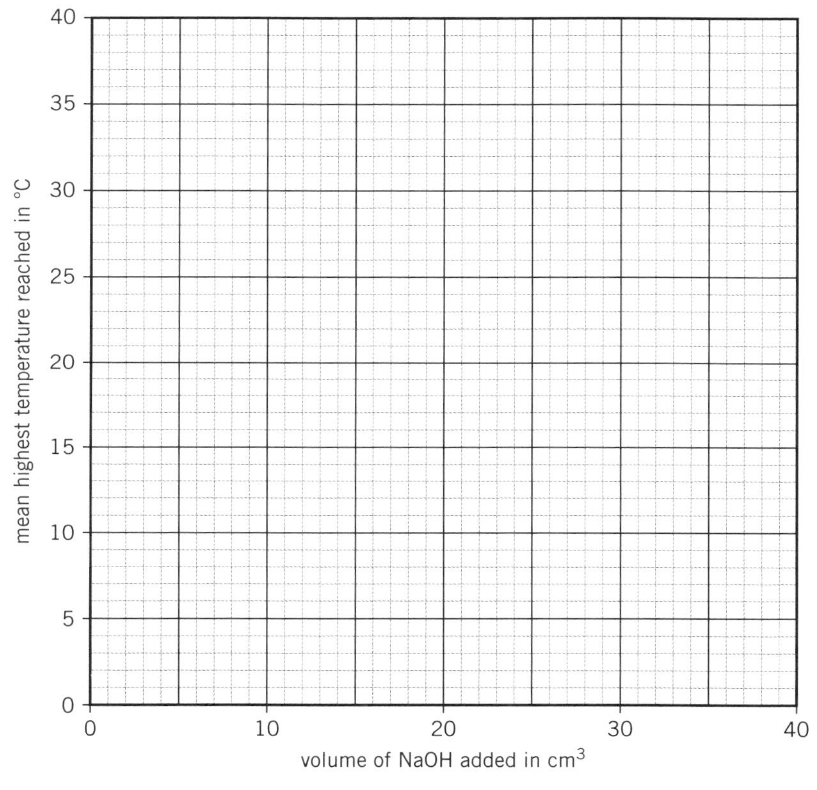

Hint

You will need to draw **two** lines of best fit on your graph – one as the temperature rises and the other as it falls.

b Estimate a value for the highest temperature reached. [2 marks]

Estimated highest temperature = ______________ °C

c Describe the shape of the graph. [2 marks]

____

____

____

____

d Explain the shape of the graph. [3 marks]

Exam Tip

Describe and explain are very different command words. These two questions may look the same, but they have different answers. Describe is what it looks like, explain is why it looks like that.

e Suggest a piece of equipment that would allow the maximum temperature to be recorded without having to draw lines of best fit on a graph. [1 mark]

11 A student added sodium hydroxide solution in 10 cm^3 intervals and used their results to estimate the highest temperature change if 23 cm^3 of sodium hydroxide solution was added.

Describe how they could change the method to confirm their estimate.

[3 marks]

12 A student is carrying out two reactions:

- copper(II) sulfate solution and iron filings
- dilute nitric acid with potassium hydroxide.

a Describe a method the student could use to investigate whether the reactions are endothermic or exothermic. Include details of what the student would expect to see for each result. [6 marks]

Exam Tip

When being asked to suggest a method for an investigation, read what the aim is carefully. In this case the student is not necessarily interested in finding the maximum temperature rise.

In a similar experiment, 28 g of iron filings react with the copper(II) sulfate solution according to the equation:

$$Fe(s) + CuSO\ \ (aq) \longrightarrow FeSO\ \ (aq) + Cu(s)$$

b Calculate the relative formula masses of copper(II) sulfate and iron sulfate. [2 marks]

____

____

____

Relative formula mass of copper(II) sulfate = ____________

Relative formula mass of iron(II) sulfate = ____________

c (H) Calculate the number of moles of iron filings that react. [2 marks]

____

____

Number of moles of iron reacting = ____________ moles

d (H) Calculate the mass of iron(II) sulfate produced in this experiment. [2 marks]

Mass of iron(II) sulfate produced = ____________ g

e (H) Explain why some reactions are exothermic and some are endothermic. [4 marks]

____

____

____

____

____

____

Hint

This question expects you to talk about covalent bonds being broken and made.

13 (H) The reaction between sodium hydroxide solution and dilute hydrochloric acid is an example of neutralisation.

Write the ionic equation for this neutralisation reaction, including state symbols. [3 marks]

____

5 Rates of reaction

Investigate how changes in concentration of a reactant affect the rate of a reaction.

Method

A Measuring the production of a gas

1. Fill the water trough and the measuring cylinder with water, and clamp the cylinder upside down in the water trough.
2. Set up the conical flask, bung, and delivery tube so that the exit of the delivery tube is under the measuring cylinder.
3. Add 50 cm^3 of 2 mol/dm^3 hydrochloric acid into the conical flask.
4. Sandpaper 3 cm of magnesium ribbon, drop it into the conical flask, quickly replace the bung, and start the stopwatch.
5. Record the volume of gas produced every 10 seconds until no more gas is being produced.
6. Repeat steps 1 to 5, for each concentration of hydrochloric acid.

B Measuring reaction rate by change in turbidity

1. Add sodium thiosulfate solution and distilled water to the conical flask in the following proportions to make each concentration:

Volume of distilled water in cm^3	40	30	20	10	0
Volume of sodium thiosulfate in cm^3	10	20	30	40	50
Final sodium thiosulfate concentration in g/dm^3	8	16	24	32	40

2. Add 10 cm^3 dilute hydrochloric acid, place the conical flask on the black cross, and start the stopwatch.
3. Record the time when the black cross is no longer visible.
4. Repeat steps 1–3 for the other concentrations of sodium thiosulfate.

Safety

- dilute hydrochloric acid: IRRITANT
- Eye protection should be worn.

Equipment

- conical flasks
- rubber bung and delivery tube to fit conical flask
- water trough
- clamp stand, boss, and clamp
- 100 cm^3 measuring cylinders
- stopwatch
- dilute hydrochloric acid at different concentrations (between 0.25 and 2.0 mol/dm^3)
- 3 cm strips of magnesium ribbon
- sodium thiosulfate solution (40 g/dm^3)
- distilled water
- printed black cross
- sandpaper

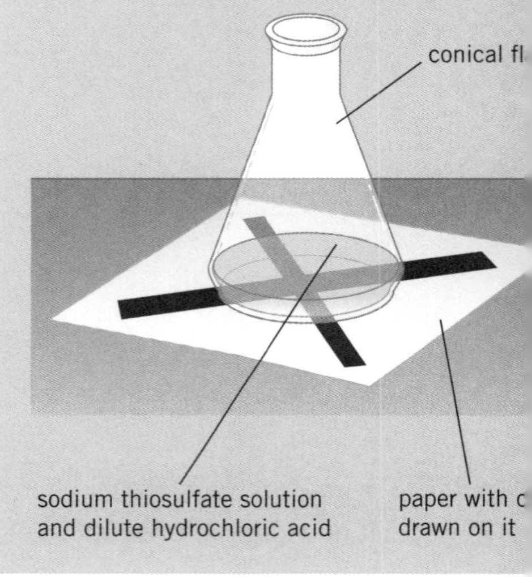

Remember !

This practical demonstrates two of the ways that the rate of a reaction can be measured. Remember that every method of measuring the rate of a reaction is actually measuring the decrease of a reactant, or the increase of a product. This could mean, for example, measuring the volume of gas forming, measuring mass lost as a solid turns to gas and escapes, or observing the formation of a coloured product.

Exam Tip 

There are many different methods that different schools will use for this practical. The important thing to remember is that the same principles are applicable across all of them.

The method for collecting gas with an inverted measuring cylinder is similar to the method for measuring the rate of photosynthesis in biology so similar questions could apply to both practicals.

This is a common practical question in exams – learn it well!

1 Give **three** ways of measuring the rate of a reaction. [3 marks]

....

....

....

2 Identify which of the following equations is correct.
Tick **one** box [1 mark]

$$\text{Mean rate of reaction} = \frac{\text{quantity of product formed}}{\text{time taken}}$$ ☐

$$\text{Mean rate of reaction} = \frac{\text{time taken}}{\text{quantity of product formed}}$$ ☐

mean rate of reaction = quantity of product formed × time taken ☐

mean rate of reaction = quantity of reactant used × time taken ☐

3 In an experiment, 56 dm^3 of carbon dioxide was produced in 15 seconds.
Calculate the mean rate of reaction.
Give your answer to 2 significant figures. [2 marks]

....

....

....

Mean rate of reaction = ... dm^3/s

4 This practical investigates the effect of reactant concentration on reaction rate. Give **three** other factors that can affect the rate of a reaction. [3 marks]

5 Explain why it is important to rub magnesium ribbon with sandpaper before using it. [3 marks]

6 During the reaction between dilute hydrochloric acid and magnesium ribbon, hydrogen gas is released.

Describe the test for hydrogen gas. [1 mark]

7 A student carried out the experiment with dilute hydrochloric acid and sodium thiosulfate solution. After the first three concentrations they lost the printed cross the teacher had provided. They decided to draw their own replacement cross and carried on, starting with the fourth concentration.

Evaluate the student's decision. [3 marks]

8 The reaction between hydrochloric acid and sodium thiosulfate can be described by the following reaction.

$$\_\_\_HCl(aq) + Na_2S_2O_3(aq) \rightarrow \_\_\_NaCl(aq) + S(s) + SO_2(aq) + H_2O(l)$$

a Complete the equation above by balancing it. [1 mark]

Hint

This equation may look much more complicated than ones you have done before but notice that there are gaps showing where you need to add in the numbers. Only put numbers in these places.

b The products all have different state symbols.

Complete the table below with:

- the name of each product
- the state symbols
- the state of the product
- any observations that you might see as the product is formed.

[5 marks]

Product	State symbol	State of product	Observation as product is made
water	(l)	liquid	no change observed
sulfur	(____)	________	________ ________
________	(aq)	________	no change observed
________	(____)	aqueous	________ ________

9 A pair of students carried out these two methods in class. They took turns doing the different jobs each time they repeated the experiment.

Explain why it is important for the same person to do the following jobs in each repeat.

a Deciding when the cross has disappeared. [2 marks]

____

____

____

b Adding the magnesium ribbon to the hydrochloric acid and putting the bung on. [2 marks]

____

____

____

10 Give the dependent variable in each of the two experiments described in the methods. [2 marks]

Dependent variable in experiment 1 = ______________________________

Dependent variable in experiment 2 = ______________________________

11 A student reacts marble chips with hydrochloric acid and collects the gas produced in a syringe. They record the volume of gas in the syringe every 5 seconds. They repeat the experiment three times using different concentrations of hydrochloric acid:

- 0.5 mol/dm^3 HCl
- 1.0 mol/dm^3 HCl
- 2.0 mol/dm^3 HCl

Predict how the results would vary for the three concentrations and explain your prediction using collision theory. [6 marks]

12 A student carried out a reaction where she kept concentration the same but changed the temperature and drew the following a graph of her results.

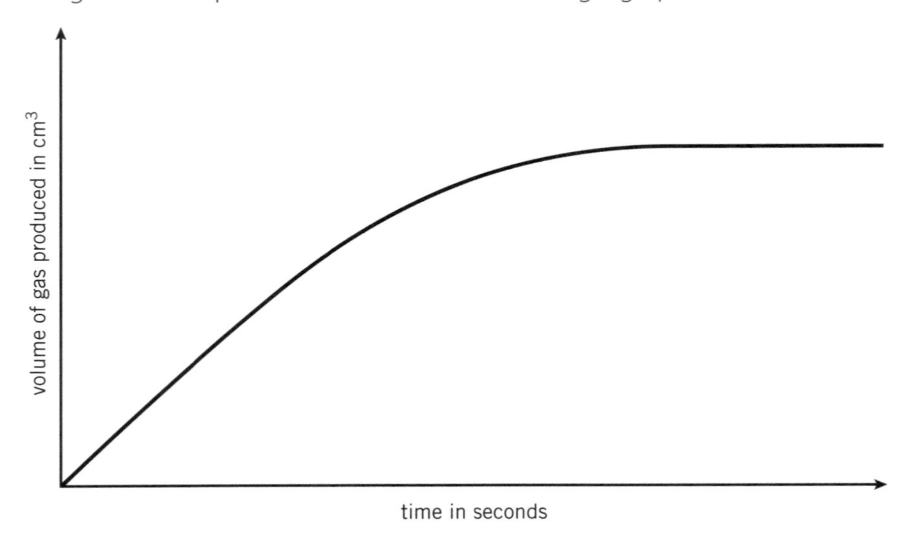

The student then repeated the experiment at a higher temperature.

a Sketch a line on the graph to show the results you would predict at a higher temperature. [2 marks]

b Explain your prediction. [3 marks]

13 A group of students carried out method **B** but could not agree exactly when the cross they were watching disappeared.

Suggest an alternative method that would reduce the amount of error in the results. [1 mark]

14 Explain why it is important to decrease the volume of water added when increasing the volume of sodium thiosulfate. [1 mark]

15 20 g sodium thiosulfate is dissolved in 0.5 dm^3 distilled water.

What is the concentration of the solution?

Tick **one** box. [1 mark]

0.5 mol/dm^3 ☐

10 g/dm^3 ☐

20 g/dm^3 ☐

40 g/dm^3 ☐

16 Marble chips reacted with hydrochloric acid in a flask. The flask was placed on a scale and the mass recorded over time.

a Plot the following results on the axes provided below. [4 marks]

Time in s	Mass of flask in g
0	95
10	43
20	27
30	21
40	20

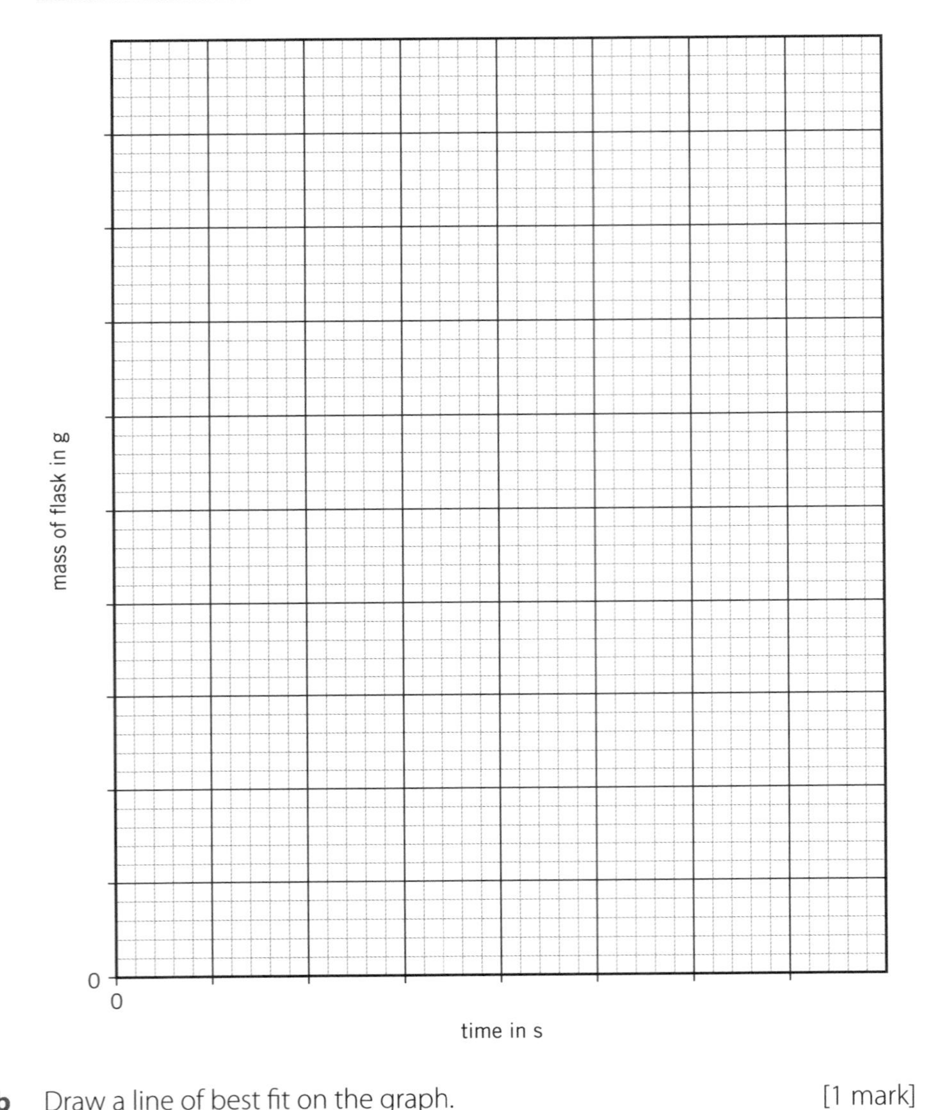

b Draw a line of best fit on the graph. [1 mark]

c Compare the rate of reaction at 10 seconds and at 30 seconds. [2 marks]

____

____

____

Exam Tip

If you are asked about the rate at a point on a curved graph, you need to draw a tangent to the curve at that point. Remember, the steeper the tangent, the faster the rate.

d **i** Write the equation to calculate the mean rate of reaction in g/s.

[1 mark]

ii Calculate the mean rate of reaction in g/s between 0 s and 40 s.

[2 marks]

____

____

Mean rate of reaction between 0 s and 40 s = _______________ g/s

e **H** Calculate the rate of reaction at 10 s:

i in g/s [2 marks]

____

____

____

Rate of reaction at 20 s = _______________ g/s

ii in mol/s [2 marks]

____

____

____

____

Rate of reaction at 20 s = _______________ mol/s

Hint

You are simply being asked to convert g/s into mol/s. This is basically the same as converting g of a substance to mol of a substance.

Think about what substance is being lost from the container. What is its relative formula mass?

6 Chromatography

Use chromatography to separate and tell the difference between coloured substances.

Method

1. Use a pencil to draw a horizontal base line, 1 cm from the bottom of the chromatography paper.
2. Use a pencil to draw a cross on the centre of the base line.
3. Use a thin paint brush or capillary tube to add some of the food colouring onto the cross and allow it to dry.
4. Fold the top edge of the chromatography paper over a wooden splint and keep in place with a paper clip.
5. Add 0.5 cm depth of water into the beaker.
6. Carefully lower the chromatography paper into the beaker, taking care to keep the pencil line above the water level. Leave until the water line (solvent front) has passed the last coloured spot.
7. Remove the chromatogram and allow it to dry.

Safety

- Do not eat the food colouring.
- Make sure you are aware of anyone with food colouring allergies.
- Capillary tubes can be fragile and care should be taken so they do not break.

Equipment

- food colourings
- capillary tubes
- chromatography paper
- pencil and ruler
- water
- 250 cm^3 beaker
- paper clip and wooden splint

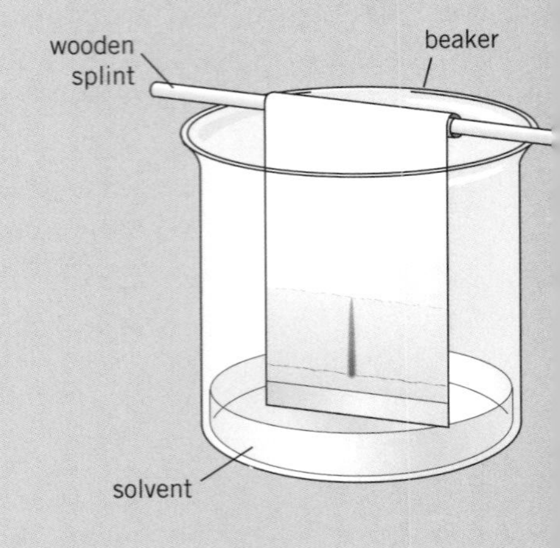

Remember !

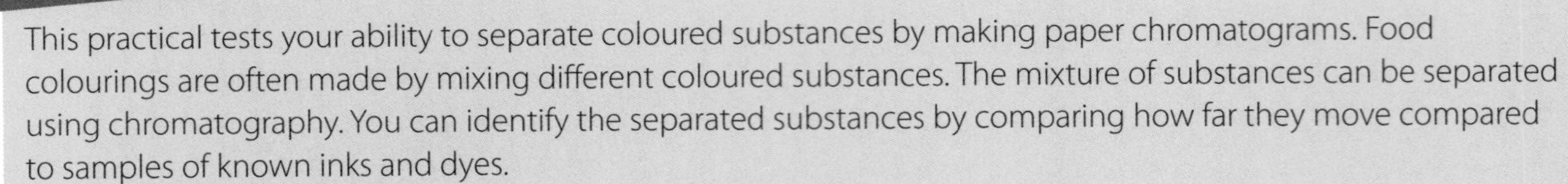

This practical tests your ability to separate coloured substances by making paper chromatograms. Food colourings are often made by mixing different coloured substances. The mixture of substances can be separated using chromatography. You can identify the separated substances by comparing how far they move compared to samples of known inks and dyes.

Exam Tip

Make sure you know how to calculate R_f values

$$R_f \text{ value} = \frac{\text{distance moved by solute}}{\text{distance moved by solvent}}$$

Remember that the solute is the spot and the liquid in the beaker is the solvent. The distance moved by solvent is the solvent front.

Chromatography involves a stationary phase and a mobile phase. In this experiment the paper is the stationary phase and the water is the mobile phase.

1 Give the meaning of the term 'mixture'. [1 mark]

____

____

2 Chromatography is one way that we can determine whether a substance is pure or not. Give two other methods can we use to determine the purity of a substance. [2 marks]

Method 1: ______________________________________

Method 2: ______________________________________

3 a Label the apparatus shown below with the terms in the box. [4 marks]

solvent	solvent front	R_f value = 0.34	R_f value = 0.60

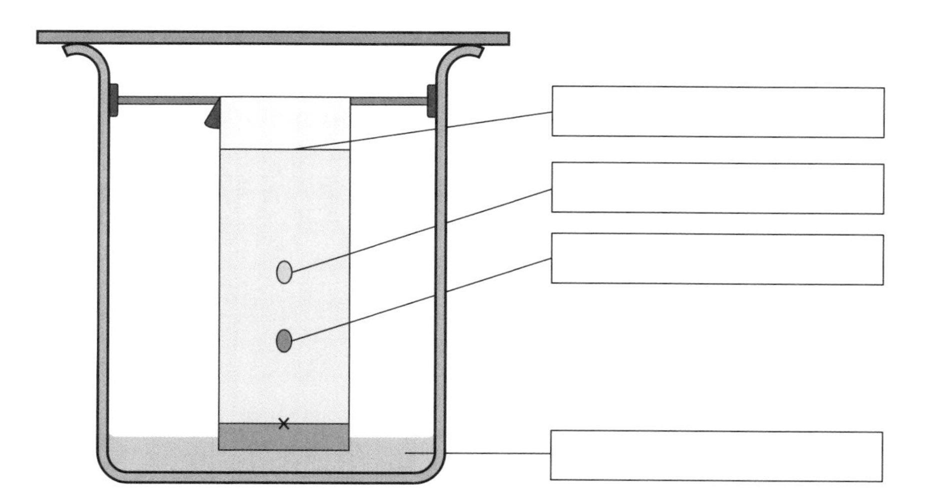

b Suggest the main purpose of the lid in the diagram in part **a**.
Tick **one** box. [1 mark]

Keeps impurities out of solvent ☐

Makes the solvent run faster ☐

Makes the solvent run slower ☐

Stops the solvent evaporating ☐

4 Chromatography can be used to test whether certain known substances are present in a sample.

Five different food colouring samples (A–E) are compared to red, blue, and yellow reference samples. The results are shown below.

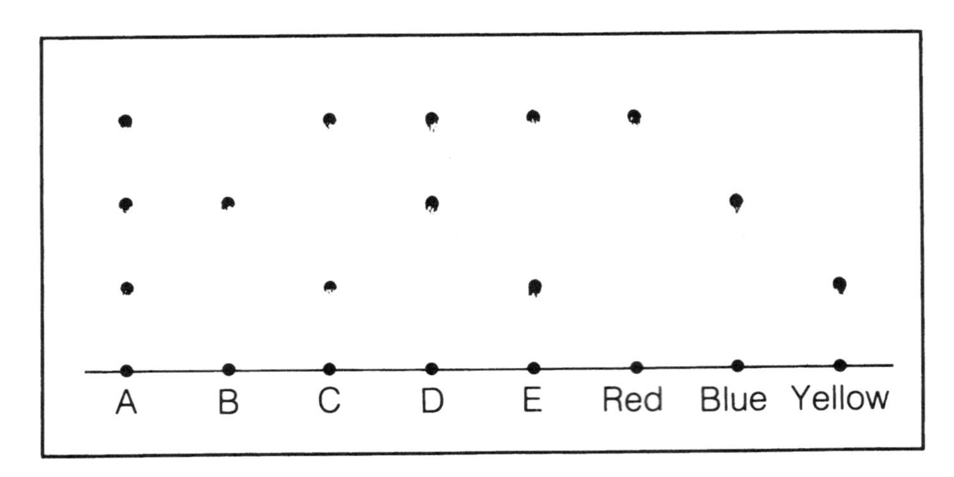

a Use evidence from the diagram to suggest which colour sample B is. [1 mark]

Colour of sample B = ____________

b How many different substances make up sample A? [1 mark]

Number of substances in sample A = ____________

c Give the letters of two unknown samples that are actually the same mixture. [2 marks]

Samples ____________ and ____________

d Use a ruler to draw a line on the diagram showing where the solvent should be in relation to the origin line at the start of the experiment. [1 mark]

e Explain why it is important to draw the origin line in pencil. [2 marks]

____

____

____

____

5 a Using the diagram below, determine which two athletes have been using banned substances. [2 marks]

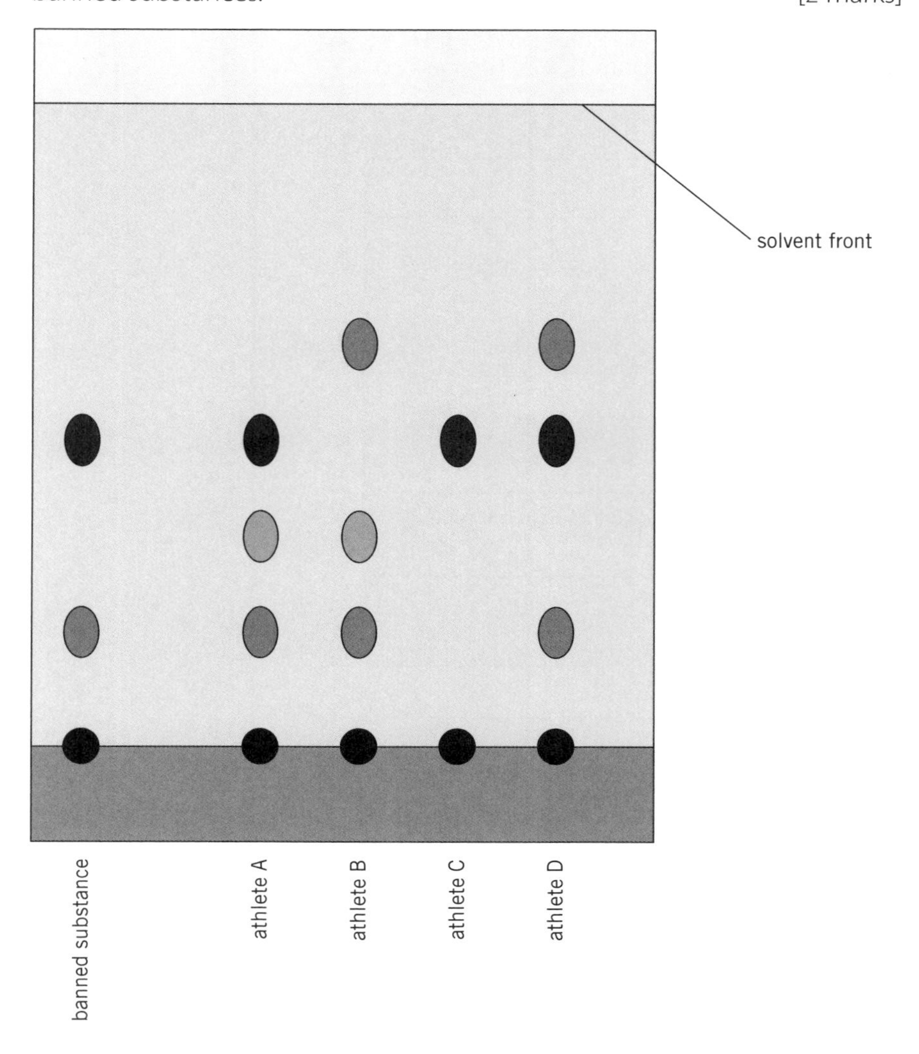

Athletes using banned substance = ________________ and ________________

b Give the equation used to calculate R_f values. [1 mark]

____

____

____

c Calculate the R_f value of the spot in sample C. [2 marks]

Hint

Make sure you measure from the middle of the spot.

____

____

____

R_f value of spot in sample C = ________________

6 Two groups of students carried out this practical.

Group A allowed solvent to move up until it had reached 75% of the way up the paper. Group B were worried about the spots spreading out too much so stopped the experiment when the solvent front was only 25% of the way up the paper.

Evaluate the methods used by the two groups. [3 marks]

7 Explain why the chromatography paper should not be allowed to touch the sides of the beaker. [2 marks]

8 Compare what you would expect to observe when a soluble sample and when an insoluble sample are spotted on the chromatography paper. [3 marks]

Exam Tip

When 'compare' is the command word, you should write about similarities and/or differences between the things you are comparing.

9 Large chemical companies can use chromatography to test their products for purity. Suggest why a company might send samples to be tested at an independent lab. [3 marks]

10 Databases of R_f values also contain information about the solvent, and the temperature used to generate the R_f values.

Explain why it is important to use the same solvent and temperature as given in the database if you want to positively identify unknown substances from a chromatogram. [4 marks]

7 Identifying ions

Use chemical tests to identify unknown compounds

Method

A Flame tests

1. Put a clean, dry nichrome wire loop into a blue Bunsen flame to ensure the loop is clean.
2. Put the clean loop into the sample mixture.
3. Put the loop back into the blue Bunsen flame and note the colour.

B Metal ion precipitation test

1. Add two drops of the solution to be tested into a dimple.
2. Then add two drops of sodium hydroxide solution.
3. Note the colour of the precipitate.

C Carbonate test

1. Half fill a test tube with limewater and put it in a test-tube rack.
2. Half fill a second test tube with the solution to be tested, and mount at a 45° angle using the stand, boss, and clamp.
3. Add 2 cm^3 dilute hydrochloric acid and quickly insert the bung and delivery tube so any gas made bubbles through the limewater.
4. Observe to see if the limewater goes cloudy.

D Sulfate test

1. Add two drops of the solution to be tested into a dimple.
2. Then add two drops of dilute hydrochloric acid and two drops of barium chloride solution.
3. If a white precipitate forms, the sample contains sulfate ions.

E Halide test

1. Add two drops of the solution to be tested into a dimple.
2. Then add two drops of nitric acid solution and two drops of silver nitrate solution. Note the colour of any precipitate formed.

Equipment

- dimple dish
- dropping pipettes
- nichrome wire loop with handle
- Bunsen burner and safety equipment
- dilute hydrochloric acid
- test tubes and rack
- bung with n-shaped delivery tube
- stand, boss, and clamp
- limewater
- dilute nitric acid
- dilute sodium hydroxide
- silver nitrate solution
- barium chloride solution
- solutions to test

Safety

- nitric acid, limewater, silver nitrate solution, sodium hydroxide, hydrochloric acid – IRRITANT
- barium chloride solution – HARMFUL
- metal salt solutions – IRRITANT / HARMFUL
- Wear eye protection and wash hands after the practical.

Remember 

This practical tests your ability to identify unknown compounds using gas tests, flame tests and precipitation reactions, some of which you will already be familiar with from other practicals.

Exam Tip 

A really common question is to ask you to identify a mystery unknown compound. I'm amazed how frequently labels fall off bottles in exam questions!

For this required practical, expect to be asked to write equations for reactions involving complicated ions. You need to learn these.

1 Complete the following table with the flame-test results you would expect to see for compounds containing each metal ion. [5 marks]

	Positive ion				
	Lithium	**Sodium**	**Potassium**	**Calcium**	**Copper**
Flame test result					

2 The table below shows a set of observations made when unknown compounds **A–C** were tested.

Unknown compound	**Qualitative test**				
	Flame test	**Metal ion precipitation test**	**Carbonate test**	**Sulfate test**	**Halide test**
A	crimson	no precipitate	limewater remains clear	No precipitate	white precipitate
B	green	blue precipitate	limewater remains clear	white precipitate	no precipitate
C	yellow	no precipitate	limewater remains clear	white precipitate formed	no precipitate

a Identify which compound is sodium sulfate. [4 marks]

Answer = ______

b Chose the correct formula for a sulfate ion.

Tick **one** box. [1 mark]

SO^{2-} ☐

SO_2^{-} ☐

SO_4^{2-} ☐

SO^{4-} ☐

c Identify the compound in sample B. [2 marks]

____

____

3 An alternative test for carbon dioxide uses a lit splint and a test tube.

a Describe how you would carry out this test and give a disadvantage of it compared with using limewater. [2 marks]

____

____

____

b Write a balanced equation for the reaction between carbon dioxide and limewater (calcium hydroxide). [2 marks]

____

____

c Explain why the limewater goes cloudy when carbon dioxide is bubbled through it. [2 marks]

____

____

4 a Iron(II) hydroxide and iron(III) hydroxide have different formulae. For each of the following compounds:

- give the formula
- describe the results of testing with sodium hydroxide.

i iron(II) hydroxide. [2 marks]

Formula = ____________

Result when tested with sodium hydroxide:

____

ii iron(III) hydroxide. [2 marks]

Formula = ____________

Result when tested with sodium hydroxide:

____

b Calcium ions, magnesium ions and aluminium ions all form a white precipitate when sodium hydroxide is added.

i Describe how you could determine if a sample contain magnesium ions or aluminium ions using sodium hydroxide. [1 mark]

____

____

ii Describe how you could differentiate between magnesium ions and calcium ions in a sample. [2 marks]

____

____

iii (H) Write a balanced ionic equation for the reaction of hydroxide ions and calcium ions. [3 marks]

Hint

This question is asking for an ionic equation so the charges on each side have to be conserved as well as the numbers of atoms balancing.

5 A student carried out flame tests on five compounds and found that all of the samples contained sodium. The rest of the class identified different positive ions for each sample, so the student decided to retest their samples.

Suggest what caused the error in the student's experiment.

Describe what the student should do to make sure the error is not repeated. [3 marks]

Exam Tip

Notice that there are two command words. Make sure you answer both parts of the question.

6 To test whether a gas is carbon dioxide (CO_2), it is bubble though limewater ($Ca(OH)_2$). If carbon dioxide is present the limewater will go cloudy.

a Write a balanced equation for the reaction between carbon dioxide and limewater (calcium hydroxide). [2 marks]

b Explain why the limewater goes cloudy. [1 mark]

7 State **three** advantages of using instrumental methods to identify the ions within a compound, instead of the methods used in this practical. [3 marks]

8 Two students tested a sample for halide ions. They could not agree about whether the precipitate was white or cream.

Describe how the students could improve their experiment and allow them to confidently interpret their results. [2 marks]

9 A student wanted to identify the negative ion in a sample.

- They tested the sample for sulfate ions in a dimple on a dimple dish. The result was negative.
- They then added nitric acid solution and silver nitrate solution to the solution in the dimple, producing a white precipitate.

The student concluded the negative ion was a halide. They were surprised to learn that the sample actually contained carbonate ions.

a Explain the error the student made and how it led to a false positive result for the halide test. [3 marks]

10 A school technician carried out tests to identify compounds that were stored in bottles whose labels had fallen off. The results are as follows.

Sample A

- Produced a gas which turned limewater cloudy when the sample was mixed with hydrochloric acid.
- White precipitate when tested with sodium hydroxide. The precipitate didn't change after excess sodium hydroxide was added.
- flame test – orange

Sample B

- White precipitate formed when mixed with hydrochloric acid and barium chloride.
- White precipitate was formed when the sample was tested with sodium hydroxide. The precipitate didn't change after excess sodium hydroxide was added.

Sample C

- flame test – crimson
- A white precipitate was formed when the sample was mixed with nitric acid and silver nitrate.

Identify the three samples. [3 marks]

____

____

____

____

Sample A = ____________

Sample B = ____________

Sample C = ____________

Exam Tip

There is a lot of information in this question. Take your time to read it carefully. You might find it useful to underline or highlight key information for each sample.

8 Water purification

Analyse and purify water samples.

Method

A Analysis

1. Use universal indicator paper to test the pH of the water sample.
2. Weigh an evaporating basin, recording the mass to 2 dp.
3. Add 5 cm^3 of the water sample to the evaporating basin and heat over a tripod and gauze until the water has evaporated.
4. Allow the evaporating basin to cool before weighing it again.
5. Subtract the original mass of the dish from the new mass to calculate the mass of dissolved, solid impurities in the water sample.

B Purification

1. Set up the apparatus as shown in the diagram in the equipment section.
2. Adjust the height of the thermometer so that the bulb is in line with the opening of the delivery tube.
3. Ignite the Bunsen burner with the air hole closed.
4. Open the air hole of the Bunsen burner so the flame turns blue and move the Bunsen burner under the tripod to heat the solution.
5. Note the temperature on the thermometer when it is at a constant value. This is the boiling point of the distillate.
6. Once half a boiling tube of distillate has been collected, remove the delivery tube and turn off the Bunsen burner.
7. Use universal indicator paper to test the pH of the distillate.

Safety

- Remember that the blue flame of the Bunsen burner is for heating and can cause burns.
- The glassware will get hot; make sure it cools before you touch it.
- Steam from the boiling mixture can cause scalds.
- Wear eye protection.

Equipment

- water sample
- Bunsen burner
- flame proof mat, tripod, and gauze
- conical flask
- two-hole bung
- universal indicator paper or pH probe
- spirit thermometer −10–110 °C
- delivery tube
- boiling tube
- large beaker with crushed ice
- anti-bumping granules
- evaporating basin
- balance

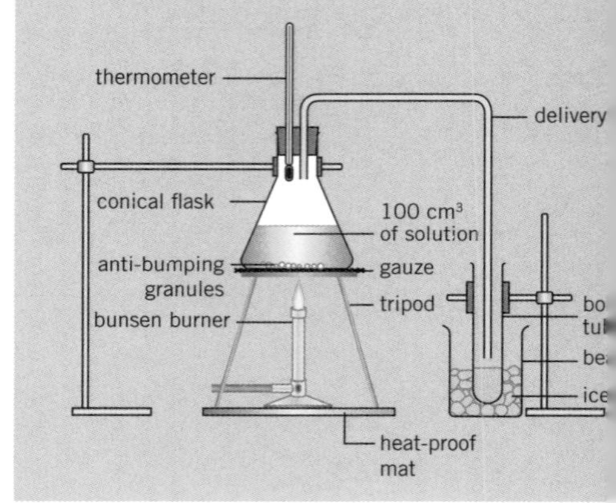

Remember

The first part of this practical is about analysing a water sample to find out how pure it is. This is done by testing the pH and working out what the mass of any solids dissolved in the water is. The second part is purifying the sample using distillation. You need to be clear about which techniques are for analysing and which are for purifying and you should be able to describe them in detail.

Exam Tip

The formula of water is H_2O. When you're writing this, the size of the number and letters is important. You will likely lose marks if you write H2O or h2o or H^2O.

1 **a** Circle the pH of pure water.

3	4	5	6	7	8	9	10	11

b Which of the following methods can be used to measure the pH of a solution?

Tick **two** boxes. [2 marks]

Blue cobalt chloride paper ☐

Boiling point ☐

Litmus paper ☐

pH probe ☐

Melting point ☐

Universal indicator solution ☐

2 Give definitions of the terms 'pure water' and 'potable water'. [2 marks]

Pure water ______________________________

____

____

Potable water ______________________________

____

____

3 Complete the table using terms from the box below. [3 marks]

desalination distillation filtration purification sterilisation

Name of process	Application
____________	Kill or remove harmful microbes.
____________	Remove insoluble particulates from sample.
____________	Separate liquids with different boiling points.

4 The diagram below shows a student's distillation experiment.

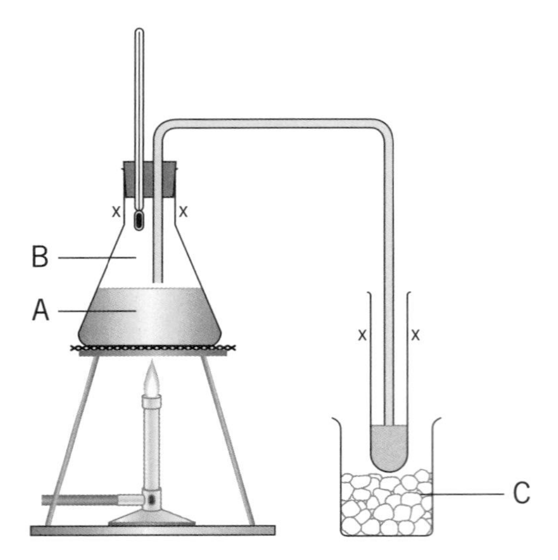

a Describe the differences in the arrangement and movement of water particles at points A, B, and C. [6 marks]

Exam Tip

Your answer doesn't always need to be all writing. A carefully annotated diagram can also gain marks in some situations.

b Use your knowledge of particle theory to explain what happens to the particles in the water sample as it is distilled. [4 marks]

Hint

This question appears to be very similar to part **a**. Make sure you really read the question and think about exactly what it is asking you. Part **a** only wanted you to describe the particles at each point. Part **b** wants you to explain the processes in terms of particle theory.

c Explain two things the student has done wrong when setting up the equipment shown in the diagram. [4 marks]

Hint

The command word is 'explain', so you should provide a reason why each thing the student has done wrong is important.

5 A student labels two solutions A and B but forgets to write down which one is which.

They know the two solutions are:

- 1 mol/dm^3 sodium chloride (NaCl)
- 1 mol/dm^3 sodium hydrogen carbonate ($NaHCO_3$)

After evaporating the water, sample A was found to have 0.585 g of salt in it and sample B had 0.84 g of salt in it.

Hint

Don't do more calculations than necessary. The solutions are the same concentration so you only need to know which solute has the greater relative formula mass.

a Identify which sample is 0.5 mol/dm^3 sodium bicarbonate and explain how you know. [3 marks]

b Suggest another way of identifying the two solutions that doesn't require the liquid to be evaporated. [1 mark]

6 The boiling point of water is 100 °C. The boiling point of ethanol is 78 °C. Both are colourless liquids.

Describe a method you could use to separate a mixture of 25% ethanol and 75% water. [6 marks]

7 A student investigates the purity of three unknown samples. Their results are shown in the table below.

	Boiling point in °C	pH	Mass of solid residue after evaporation of 100 ml solution in g	Conducts electricity?
distilled water	100.0	7.0	0.00	No
sample A	100.5	8.1	2.91	Yes
sample B	102.2	7.01	3.82	Yes
sample C	98.3	7.24	0.00	No

a Identify which sample has the most impurities dissolved in it and give a reason for your choice. [2 marks]

Sample with most dissolved impurities = ______

Reason

b The student says that measuring pH alone is not enough to measure the purity of a water sample.

Give evidence from the table that supports this statement. [2 marks]

c Sample C has no solid residue left when evaporated.
Suggest why it has a boiling point 1.7 °C lower than distilled water. [1 mark]

____

____

Hint

Look at the number of marks available for each question. This is only worth one mark so it is only asking for a simple answer – not a whole paragraph.

d The solid residue in samples A and B is found to be sodium chloride.
Explain why samples A and B conduct electricity, but distilled water does not. [2 marks]

____

____

____

____

8 Describe what happens to blue cobalt chloride paper when it is exposed to water. [1 mark]

____

____

Periodic table

key

relative atomic mass
atomic symbol
name
atomic (proton) number

1	2											3	4	5	6	7	0
							1 **H** hydrogen 1										4 **He** helium 2
7 **Li** lithium 3	9 **Be** beryllium 4											11 **B** boron 5	12 **C** carbon 6	14 **N** nitrogen 7	16 **O** oxygen 8	19 **F** fluorine 9	20 **Ne** neon 10
23 **Na** sodium 11	24 **Mg** magnesium 12											27 **Al** aluminium 13	28 **Si** silicon 14	31 **P** phosphorus 15	32 **S** sulfur 16	35.5 **Cl** chlorine 17	40 **Ar** argon 18
39 **K** potassium 19	40 **Ca** calcium 20	45 **Sc** scandium 21	48 **Ti** titanium 22	51 **V** vanadium 23	52 **Cr** chromium 24	55 **Mn** manganese 25	56 **Fe** iron 26	59 **Co** cobalt 27	59 **Ni** nickel 28	63.5 **Cu** copper 29	65 **Zn** zinc 30	70 **Ga** gallium 31	73 **Ge** germanium 32	75 **As** arsenic 33	79 **Se** selenium 34	80 **Br** bromine 35	84 **Kr** krypton 36
85 **Rb** rubidium 37	88 **Sr** strontium 38	89 **Y** yttrium 39	91 **Zr** zirconium 40	93 **Nb** niobium 41	96 **Mo** molybdenum 42	[98] **Tc** technetium 43	101 **Ru** ruthenium 44	103 **Rh** rhodium 45	106 **Pd** palladium 46	108 **Ag** silver 47	112 **Cd** cadmium 48	115 **In** indium 49	119 **Sn** tin 50	122 **Sb** antimony 51	128 **Te** tellurium 52	127 **I** iodine 53	131 **Xe** xenon 54
133 **Cs** caesium 55	137 **Ba** barium 56	139 **La*** lanthanum 57	178 **Hf** hafnium 72	181 **Ta** tantalum 73	184 **W** tungsten 74	186 **Re** rhenium 75	190 **Os** osmium 76	192 **Ir** iridium 77	195 **Pt** platinum 78	197 **Au** gold 79	201 **Hg** mercury 80	204 **Tl** thallium 81	207 **Pb** lead 82	209 **Bi** bismuth 83	[209] **Po** polonium 84	[210] **At** astatine 85	[222] **Rn** radon 86
[223] **Fr** francium 87	[226] **Ra** radium 88	[227] **Ac*** actinium 89	[261] **Rf** rutherfordium 104	[262] **Db** dubnium 105	[266] **Sg** seaborgium 106	[264] **Bh** bohrium 107	[277] **Hs** hassium 108	[268] **Mt** meitnerium 109	[271] **Ds** darmstadtium 110	[272] **Rg** roentgenium 111	[285] **Cn** copernicium 112	[286] **Nh** nihonium 113	[289] **Fl** flerovium 114	[289] **Mc** moscovium 115	[293] **Lv** livermorium 116	[294] **Ts** tennessine 117	[294] **Og** oganesson 118

*The lanthanides (atomic numbers 58–71) and the actinides (atomic numbers 90–103) have been omitted.

Relative atomic masses for **Cu** and **Cl** have not been rounded to the nearest whole number.

Notes – 1 Making salts

Make notes on the practical you carried out.

Hint

Make notes on:
- your method
- safety precautions
- sources of error
- possible improvements
- the function of equipment
- how to make accurate measurements.

Tip

You could use this space to:
- draw a labelled diagrams showing filtration, evaporation, and crystallisation
- write the general acid equations.

Notes – 2 Neutralisation

Make notes on the practical you carried out.

Hint

Make notes on:
- your method
- safety precautions
- sources of error
- possible improvements
- the function of equipment
- how to make accurate measurements.

Tip

You could use this space to:
- draw a labelled diagram of the titration experiment
- write the equation linking volume, concentration, and mass.
- (H) write the equation linking mass, relative formula mass, and number of moles.

Notes – 3 Electrolysis

Make notes on the practical you carried out.

Hint

Make notes on:
- your method
- safety precautions
- sources of error
- possible improvements
- the function of equipment
- how to make accurate measurements.

Tip

You could use this space to draw a labelled diagram of the equipment used to electrolyse an ionic solution.

Notes – 4 Temperature changes

Make notes on the practical you carried out.

Hint

Make notes on:
- your method
- safety precautions
- sources of error
- possible improvements
- the function of equipment
- how to make accurate measurements.

Tip

You could use this space to sketch reaction profiles of:
- an endothermic reaction
- an exothermic reaction.

Notes – 5 Rates of reaction

Make notes on the practical you carried out.

Hint

Make notes on:
- your method
- safety precautions
- sources of error
- possible improvements
- the function of equipment
- how to make accurate measurements.

Tip

You could use this space to:
- write the equation linking rate of reaction, time, and mass of substance formed
- sketch a graph of volume of gas produced against time.

Notes – 6 Chromatography

Make notes on the practical you carried out.

Hint

Make notes on:
- your method
- safety precautions
- sources of error
- possible improvements
- the function of equipment
- how to make accurate measurements.

Tip

You could use this space to:
- write the equation for R_f value
- draw a labelled diagram of a chromatography experiment.

Notes – 7 Identifying ions

Make notes on the practical you carried out.

Hint

Make notes on:
- your method
- safety precautions
- sources of error
- possible improvements
- the function of equipment
- how to make accurate measurements.

Tip

You could use this space to draw a table of the positive results from the five tests for ions.

Notes – 8 Water purification

Make notes on the practical you carried out.

Hint

Make notes on:
- your method
- safety precautions
- sources of error
- possible improvements
- the function of equipment
- how to make accurate measurements.

Tip

You could use this space to draw a labelled diagram of the distillation equipment.

Notes

Great Clarendon Street, Oxford, OX2 6DP, United Kingdom

Oxford University Press is a department of the University of Oxford. It furthers the University's objective of excellence in research, scholarship, and education by publishing worldwide. Oxford is a registered trade mark of Oxford University Press in the UK and in certain other countries

First published in 2019

British Library Cataloguing in Publication Data
Data available

978 0 19 844491 6

10 9 8 7 6 5 4 3 2 1

Paper used in the production of this book is a natural, recyclable product made from wood grown in sustainable forests.
The manufacturing process conforms to the environmental regulations of the country of origin.

Printed in Great Britain by Bell and Bain Ltd. Glasgow

Acknowledgements

COVER: CHARLES D. WINTERS/SCIENCE PHOTO LIBRARY

Artwork by Aptara Inc.